Understanding the Elements of the Periodic Table™

COBALT

Paula Johanson

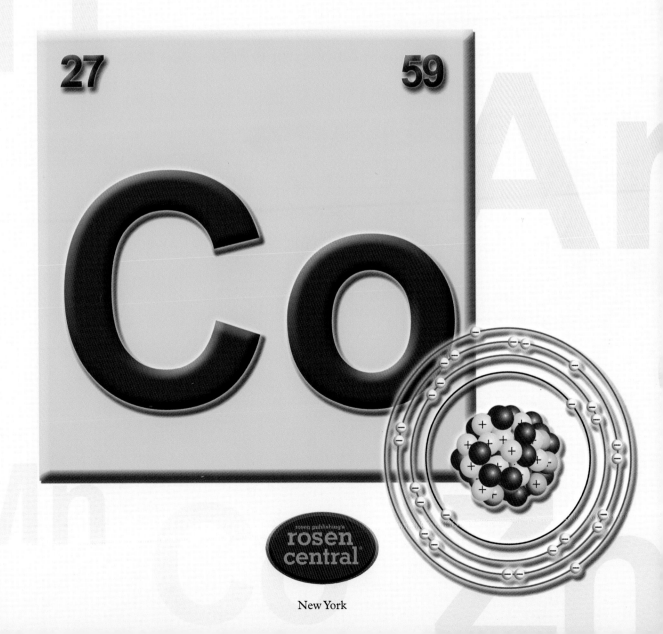

27   59

Co

rosen publishing's
rosen central®

New York

To the librarians of our school and public libraries

Published in 2008 by The Rosen Publishing Group, Inc.
29 East 21st Street, New York, NY 10010

Library of Congress Cataloging-in-Publication Data

Johanson, Paula.
Cobalt / Paula Johanson.—1st ed.
    p. cm.—(Understanding the elements of the periodic table)
Includes bibliographical references and index.
ISBN-13: 978-1-4042-1410-1 (hardcover)
1. Cobalt—Juvenile literature. 2. Chemical elements—Juvenile literature.
I. Title.
QD181.C6J64 2008
546'.623—dc22

                                                                    2007023665

Manufactured in Malaysia

On the cover: Cobalt's square on the periodic table of elements; *(inset)* model of the subatomic structure of a cobalt atom.

Contents

# Introduction

Before the discovery of cobalt (chemical symbol: Co) in 1735, there were only a few metals known for thousands of years. As people developed mines to dig up ores, miners and smiths discovered that some of what they dug up didn't actually produce metal when smelted, or melted. All that was produced was a horrible garlicky smell and crumbly rock. Was this some kind of joke? The ancient Greek word for "mine" was *kobalos*, from the Greek words for "house-spirit" and "old." This word eventually became the German word *kobold* and the English word "goblin." Miners wondered if there might be spirits living in mines that led them to find metal ores and sometimes played tricks.

It did seem worthwhile, though, to keep looking for gold (Au) and other valuable substances. At the same time, scholars made dedicated studies of the natural world. The alchemists kept looking for the fabled philosopher's stone in particular. They believed it could change other metals into gold, give eternal life, and lead to spiritual enlightenment. Though they never found the philosopher's stone, what they learned from their searching eventually became the science of chemistry. Scholars and scientists gradually learned that instead of only a few elements, there were many elements to be found. Among the first to be separated out from ores was a new metal. The discoverer of this new metal, Swedish chemist Georg Brandt (1694–1768), named it cobalt in 1735 after the mine spirits.

You won't often see pieces of pure cobalt like these outside of a laboratory or a chemistry display. In nature, cobalt is found as compounds in rocky ores. In metallic meteorites, there is iron mixed with nickel and a little cobalt.

The study of metals and their chemistry is called metallurgy. Metal workers have been using familiar metals such as iron (Fe), copper (Cu), silver (Ag), and gold for thousands of years. However, cobalt was a new metal in the 1700s. The discovery of cobalt as a new element was an important step in the emerging modern science of metallurgy.

Chapter One
The Element Cobalt

The element cobalt was the first metal to be discovered by modern scientific methods. Cobalt, which is represented by the chemical symbol Co, is a metal rarely found in its pure state in nature. But cobalt compounds have been used in pottery glazing, glassmaking, and enamel work since ancient times. Pure cobalt is a shiny metal with a silvery color resembling iron. Cobalt is often combined with iron and other metals to make useful alloys for the manufacture of very hard tools or permanent magnets.

Atoms, Elements, and Compounds

The world and everything you see in it—and nearly everything in the universe, no matter how far away—is made up of elements. Some objects are made of one element only. Other objects have several elements bonded chemically into compounds.

Just how small a piece of an object can you get? If you started with a teapot, you could see that it isn't all one substance. It's made from hard clay with some shiny blue and white glaze on the outside. The smallest piece of blue glaze you could cut would have some cobalt in it, along with the other elements silicon (Si) and oxygen (O). You could use chemical reactions to separate the silicon and oxygen and leave only the cobalt.

This teapot was glazed with cobalt compounds to make it a strong, bright blue color. Cobalt is used in pottery glazes around the world.

And the tiniest piece that you could get of the cobalt is an atom. Atoms are far too small to see with a microscope. Atoms are so tiny that 20 million cobalt atoms side by side would make a line only 0.4 of an inch (1 centimeter) long.

Inside the Atom

The smallest piece of cobalt that you could ever have would be one atom. But inside that atom are even smaller pieces called subatomic particles. There are three main subatomic particles found inside each atom: protons, neutrons, and electrons. At the center of the atom is a dense core made of protons, which carry a positive electrical charge, and neutrons, which have no charge. This core is called the nucleus, and it contains nearly all of the mass in an atom. Cobalt has twenty-seven protons in its nucleus. That is what makes it an atom of cobalt. If an atom anywhere in the universe contains twenty-seven protons, it is an atom of cobalt.

The electrons in an atom orbit the atom's nucleus in layers called shells. The negative electrical charge of the electrons attracts them to the positive protons in the nucleus. In a neutral atom, the negative and positive electrical charges of the atom are balanced. The number of electrons and protons are equal. An atom of cobalt has twenty-seven protons and twenty-seven electrons.

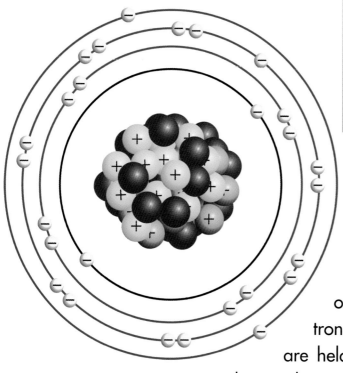

This illustration shows electrons close to the nucleus of a cobalt atom. A realistic model would have tiny electrons zooming around a stadium, with a raspberry at the center for the nucleus.

The outermost shell of electrons is a little less tightly held to the nucleus than the inner shells. The electrons in the outermost shell are valence electrons. In metals, the valence electrons are held so weakly that they can leave the metal atom and enter a different atom.

The Periodic Table

Although they were not recognized as such, some elements like carbon (C), copper, gold, iron, and mercury (Hg) have been known since ancient times. After they were identified as elements, and as more elements were discovered and studied over the years, differences and similarities between them were recognized. When elements are arranged in order of increasing atomic number (the atom's number of protons), the elements show similar patterns in properties. The patterns tend to repeat periodically. These patterns can be highlighted by arranging the elements on a chart called the periodic table of elements. The most common form of periodic table was invented in 1869, by a Russian chemist named Dmitry Mendeleyev. He used it to help teach his students at the University of St. Petersburg in Russia.

H																	He
Li	Be											B	C	N	O	F	Ne
Na	Mg											Al	Si	P	S	Cl	Ar
K	Ca	Sc	Ti	V	Cr	Mn	Fe	Co	Ni	Cu	Zn	Ga	Ge	As	Se	Br	Kr
Rb	Sr	Y	Zr	Nb	Mo	Tc	Ru	Rh	Pd	Ag	Cd	In	Sn	Sb	Te	I	Xe
Cs	Ba	La	Hf	Ta	W	Re	Os	Ir	Pt	Au	Hg	Tl	Pb	Bi	Po	At	Rn
Fr	Ra	Ac	Rf	Db	Sg	Bh	Hs	Mt	Uun	Uuu	Uub		Uuq		Uuh		Uuo

	Ce	Pr	Nd	Pm	Sm	Eu	Gd	Tb	Dy	Ho	Er	Tm	Yb	Lu
	Th	Pa	U	Np	Pu	Am	Cm	Bk	Cf	Es	Fm	Md	No	Lr

Each element's place on the periodic table shows its chemical symbol. Cobalt (Co) is located among the metals, in a section for transition metals. The transition metals are somewhat less reactive than the alkali metals on the far left of the table.

An obvious pattern on the periodic table is the difference between the metals, which are on the left side of the chart, and the nonmetals, which are on the right side of the chart. (See the periodic table on pages 40–41.)

Most metals have similar qualities, such as the ability to conduct electricity. They can be polished to be shiny. Most metals are malleable, which means they can be pounded into shapes without breaking. Malleable metals are usually ductile, which means they can be pulled into wires.

The nonmetals are grouped together to the right of a "staircase" line across the chart. Among these nonmetals are the noble gases, which do not react with any elements under most conditions, on the far right of the chart. Also among the nonmetals are oxygen and nitrogen (N), the major gases in the air, and sulfur (S) and phosphorus (P), two solids used in making matches.

Some elements have a few properties in common with metals and other properties like nonmetals. These elements are called metalloids. Metalloids such as boron (B) and silicon are brittle, black solids, but they conduct electricity, though less well than metals. They are located on that staircase line, between metals and nonmetals.

Cobalt is located on the left of that staircase line, so it is among the metals. The most reactive metals are those on the far left of the periodic table, the alkali metals, which include sodium (Na) and potassium (K). Cobalt is farther to the right and is not as reactive as the alkali metals. Cobalt is in a section of elements called the transition metals. They are called transition metals because they form a transition between the most reactive metals on the left of the periodic table and the nonmetals on the right.

Periods and Groups

Each row of elements on the periodic table is called a period. The periods are numbered down from the top, and the numbers indicate the number of electron shells in an atom of each element. Because cobalt is in period 4, it has four shells of electrons. As you read across the row, each atom has one more proton in its nucleus than the element on its left and also one more electron in its outer shell.

On the periodic table, the columns of elements are called groups. Elements in a group have similar properties. Cobalt is in group VIIIB (also called group 9) with rhodium (Rh) and iridium (Ir). All the elements in group VIIIB are in the ninth column from the left side of the chart. As you read down the column, cobalt has four electron shells, rhodium has five, and iridium has six.

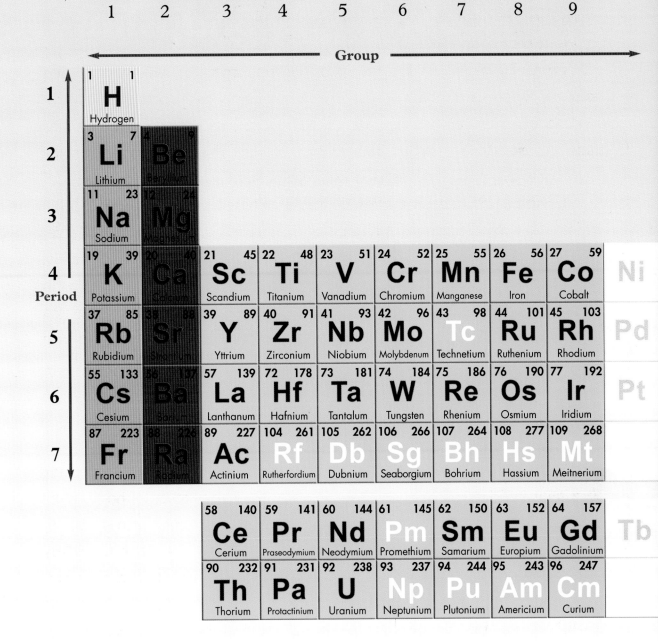

In the periodic table, each element in a group has similar properties. Each element in a period has one more proton in its atom's nucleus than that of the element on its left and one proton fewer than the nucleus of the element on its right. Cobalt is located in period 4 and group 9.

Cobalt Snapshot

Chemical Symbol:	Co
Classification:	Transition metal, group VIIIB (9)
Properties:	Silvery color, brittle, hard metal that is attracted by a magnet
Discovered by:	Georg Brandt in 1735
Atomic Number:	27
Atomic Weight:	58.933 atomic mass units (amu)
Protons:	27
Electrons:	27
Neutrons:	31 or 32
State of Matter at 68° Fahrenheit (20° Celsius):	Solid
Melting Point:	2,723°F (1,495°C)
Boiling Point:	5,301°F (2,927°C)
Commonly Found:	In Earth's crust as cobalt ores; also in metallic meteorites

Unique Qualities of Elements

The number of protons in the nucleus of an atom defines the element. On the periodic table, you can see the atomic number 27 written above and left of the symbol Co for cobalt. An atom with twenty-six protons would be a different element with different properties, and that element is iron. An atom with twenty-eight protons is nickel (Ni). Just one proton more or less makes all the difference to make nickel shiny and silvery, cobalt shiny and hard, and iron gray and more malleable.

Atomic Weight

On the periodic table, above and to the right of the symbol Co for cobalt, you can see the number 59. That is the approximate atomic weight of cobalt: the number of protons plus the average number of neutrons found in an atom of that element. Atomic weight is also known as atomic mass, and it is expressed in atomic mass units (amu). The atomic weight of cobalt is 58.933 amu, which has been rounded to two digits, 59, for our periodic table. (See pages 40–41.)

Isotopes

Cobalt atoms do not always have the same number of neutrons in the nucleus. The different forms of an element with differing numbers of neutrons are called isotopes. All cobalt atoms found in nature on Earth have thirty-two neutrons. Scientists have made other isotopes, but they are unstable and radioactive.

Many elements can come in different forms, or allotropes. Allotropes of an element can appear as different as graphite and diamonds. Both contain only carbon atoms, but one carbon allotrope is gray and slippery, while the other is clear and hard. Pure cobalt metal also has allotropes, but most cobalt metal is a mixture of two allotropes. As the temperature changes, cobalt can change from one allotrope to the other, but the transformation is sluggish. This accounts in part for the variation in the physical properties of cobalt.

Transition Metals

When you look across a row of the periodic table from left to right, you are looking at a period—a series of elements with an increasing number of electrons. The element on the right end of the period is a nonmetal with a filled outermost shell of electrons, and there is no room for more electrons. Therefore, it doesn't react with other elements. The element that is second from the right has an outermost shell that has room for one more electron. This element reacts with elements that can easily lose electrons. The element on the left end of the period in group 1 has just one electron in its outermost shell, and it can lose this electron very easily to become an ion. Because it can easily lose an electron, it is very reactive. The elements in the middle part of the period have outermost electrons shells that are only partly full. These metals aren't as reactive as metals in group 1. But they're definitely not nonmetals that want to take up a spare electron to fill their electron shells. These elements in the middle of the periodic table are called transition metals.

Cobalt's Hardness

Pure cobalt is a moderately hard metal. We express the hardness of metals, minerals, and other materials with Mohs' scale. The German mineralogist Friedrich Mohs published his scale in 1822. From miners, he learned the idea of using scratch tests to compare minerals. His scale compares ten groups of substances from softest to hardest. From 1 to 2.5 on his scale are materials considered soft. Between 2.5 and 5.5 are substances of an intermediate hardness. Above 5.5, materials are considered hard. Diamond, the hardest substance, is in the tenth group. On the Mohs' scale, cobalt has a hardness of 5. Your fingernail has a hardness of 2.5 on this scale, and a piece of common quartz rock has a hardness of 7. A tool made of cobalt could scratch your fingernail, but a piece of quartz could scratch the cobalt.

Heat-Sensitive Cobalt Invisible Ink

Have you ever written a secret message in lemon juice? When the juice dries on the paper, it's almost invisible. But if you heat the paper, the juice turns brown, and your secret message can be read.

Cobalt compounds make good invisible inks. If the paper is heated carefully and doesn't get too hot, the secret message will appear and then fade as the paper cools. There are even different colored inks, depending on which cobalt compound is used. For a while in the 1800s, fireplace screens and novelty pictures were drawn using ordinary ink for part of the

Mohs' Scale

Hardness Rating Examples

1 Talc

2 Gypsum (and rock salt, fingernails)

3 Calcite (copper)

4 Fluorite (and iron)

5 Apatite (and cobalt)

6 Orthoclase (and rhodium, silicon,
    tungsten [W])

7 Quartz

8 Topaz (and chromium [Cr], hardened steel)

9 Corundum (sapphire or ruby)

10 Diamond

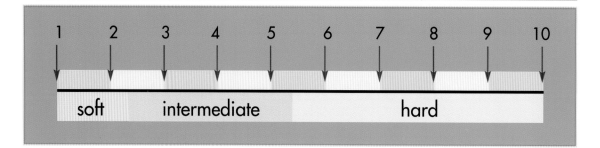

picture and heat-sensitive invisible ink for other parts. The ordinary ink could show a plain winter scene with bare trees. However, when the picture was placed near a fire, the invisible ink would become visible, putting green leaves on the trees and green grass and blooming flowers on the hills. When the screen cooled down, the colors would disappear.

Writings of this kind were a parlor trick in the 1800s and early 1900s. They were done in ordinary ink and invisible ink made from a cobalt compound that shows up in colors when heated with a lamp.

Ferromagnetic Metals

Cobalt is one of the three ferromagnetic metals—iron, nickel, and cobalt—that are the only pure metals that can become permanent magnets. In other words, cobalt becomes magnetized in a magnetic field and keeps its magnetism, even after the field has been removed. Cobalt has a magnetic permeability about two-thirds that of iron: it lets a magnetic field pass through it pretty easily. Most other metals do not let much of a magnetic field through them, and most can't be picked up by a magnet. Some substances, such as wood and most plastics, don't cling to a magnet at all.

It can be tedious work to sort through piles of used metal for recycling. But in North America and Europe, electromagnets are used to speed up the process of sorting. Cobalt is a ferromagnetic metal, and it can become a permanent magnet once it has been in the coil of an electromagnet.

An electromagnet is a device that creates a magnetic field using electricity. Most electromagnets have a coil of wire, and if an electric current is run through the wire, a magnetic field is produced. If a piece of metal is put into the center of the coil, it can concentrate the magnetic field and make it stronger. However, for most metals, after the current is turned off, the piece of metal is no longer a magnet. Only cobalt, nickel, and iron (and some of their alloys) can become a permanent magnet after being in the coil of an electromagnet.

If you've put a magnet on your refrigerator door, you may have handled a magnet made from a cobalt alloy. One alloy with iron, aluminum (Al), nickel, and cobalt, called alnico, is commonly used to make small strong magnets.

Cobalt in Magnetic Data Recording

Until recently, if you made a sound recording or a video recording, you would have used audiotapes and videotapes. These tapes are still in use today. The tape is a thin plastic film coated with magnetic iron oxides with

Sorting Metal Trash

Junkyards can have piles of crushed cars and broken machinery. There, workers separate the various kinds of metals to be used again. But how do they tell which is which? One way to make the sorting easier is to put an electromagnet onto the end of a crane. Turn on the electric current, and then only magnetic metal pieces will cling to the end of the crane. Swing the crane arm over an empty bin and turn off the electric current. At least some of the metal scraps have been sorted quickly!

a small percentage of cobalt added for high-density recording. There are similar magnetic strips on credit cards and bank cards. Don't bring a magnet near a cassette tape or your wallet! The magnetic recordings could be wiped out.

Radioactive Isotopes

Radioactivity is an important quality of a special isotope of cobalt, called cobalt-60. This isotope is artificially created by bombarding natural cobalt (cobalt-59) with neutrons. It was discovered in 1938 by the scientists

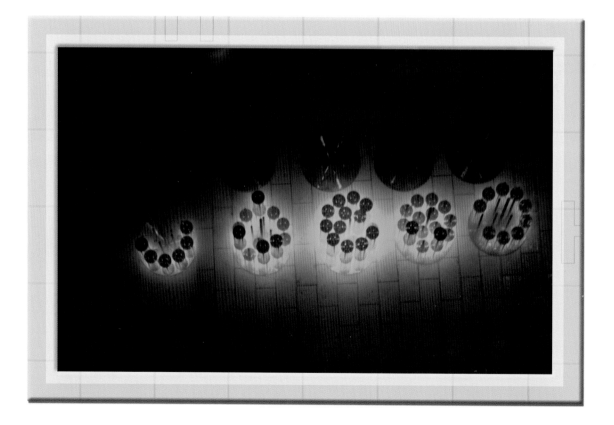

These samples of the radioactive isotope cobalt-60 are making the water glow blue with Cerenkov radiation, emitted when charged particles travel through a medium such as water. This is the only way you'll actually see this kind of radiation.

Glenn Seaborg and John Livingood. Cobalt-60 has 27 protons and 33 neutrons, and it is unstable. It has a half-life of 5.27 years. That means that after 5.27 years, half of the cobalt-60 atoms will have decayed to a stable atom, nickel-60, releasing radiation. This radiation makes this isotope very useful because the radiation will show on X-ray photographs or radiation meters. It can be used as a tracer in industry to show where pipes are leaking.

Doctors use radioactive isotopes of cobalt as a radioactive dye to trace out blood vessels in a person's body. A person's veins and arteries are not visible on an X-ray photograph. But with a little radioactive dye, the blood vessels emit radiation that can be detected and studied by computerized axial tomography, or a CAT scan. Many types of illness cause a change in the way the arteries grow. Radioactive dye is very useful for diagnosis. Only a small amount of the isotope is used, not enough to cause radiation poisoning.

Radioactive cobalt is also used to treat cancer. Some types of cancer cells are very easy to kill with a low dose of radiation that doesn't hurt healthy body cells. Doctors can use a controlled dose of radiation aimed exactly at a tumor.

Chapter Three
Where Can Cobalt Be Found?

Cobalt is the thirtieth most abundant element in the rocks of Earth's crust; it is just a little more common than copper. It doesn't occur in veins of pure metal, such as gold or silver. It is found as compounds in volcanic rock and ore. The major commercial sources for cobalt are mines in Ontario, Canada, which have cobaltite ores, and mines in Zambia, Morocco, and the Democratic Republic of the Congo in Africa.

Scientists know from studying the light of stars that there is cobalt elsewhere in the solar system and the universe. Out of every ten million atoms in the universe, less than one is cobalt. Most of the atoms in the universe are hydrogen (H) and helium (He) in big clouds of gas and in stars. All the cobalt in the universe was formed inside old stars and was released when they exploded as supernovas. After the explosion, clouds containing cobalt and other elements gradually condensed to form the cores of planets and asteroids.

Discovery of Cobalt

In the sixteenth century, on the borders of Saxony and Bohemia in what is now Germany, the silver mines were producing less silver each year. Superstitious miners blamed mischievous mine spirits, which they called

In this photograph, a man cleans bags of cobalt that have been taken out of a mine in the Democratic Republic of the Congo. Since ancient times, miners working with cobalt ores have suffered lung diseases and rashes. Safety equipment and masks are needed to save lives.

kobolds, for stealing silver and leaving behind a mineral useless to silver miners. They disliked both the trouble of removing it and the dangerous arsenic ores usually found with it. Christoph Schürer found that when he added some of this mineral to molten glass or to pottery glaze, it produced a more intense blue than could be gotten from copper, which he used earlier.

The person who found a new element in this mineral was Swedish scientist Georg Brandt. Brandt helped his father in operating a copper smelter and ironworks. After studying chemistry and medicine in Leiden and Reims, Brandt returned to Stockholm to take charge of the Bureau of Mines laboratory. He worked for years with minerals that stained glass blue.

At the time, it was commonly believed that an element called bismuth (Bi) was responsible. But Brandt worked to isolate what he believed was an unknown metal.

Discovering an element is seldom as simple as picking up a chunk of gold from a streambed. Some scientists worked for years to find ways to isolate pure substances from rocks that appeared to be a mixture of substances. When Brandt heated his mineral, it released gases that smelled garlicky and horrible because the mineral contained arsenic (As). Rainwater that soaked through mine tailings would run off into puddles. Animals that drank from these puddles would become sick or die. Rainwater is slightly acidic, so Brandt tried soaking the mineral in an acid solution. He was able to remove the unknown metal from the acid solution and study it between 1735 and 1739.

At the time, there were six metals known (gold, silver, copper, iron, lead [Pb], and tin [Sn]), as well as what Brandt called "half-metals": mercury, bismuth, zinc (Zn), antimony (Sb), and arsenic. Brandt found six ways to distinguish between bismuth and the unknown metal. He called the new metal cobalt. For a couple of years, he worked with this metal before formally announcing the results of his experiments in 1741.

Cobalt Ores

Cobalt is found in several different ores. In an ore called smaltite or speiss cobalt, it is combined with

Many minerals such as this cobaltite look much like other ores. Some fool's gold was called Nick's copper or goblin's ore by miners, but it is now known as nickel or cobalt.

arsenic in the compound cobalt arsenide, $CoAs_2$. In the ore cobaltite, it is combined with arsenic and sulfur (S) in the form of cobalt sulfarsenide (CoAsS). Another ore contains the hydrated arsenate of cobalt, $Co(AsO_4)_2 \bullet 8H_2O$, and it is called erythrite or cobalt bloom.

Cobalt in Meteorites and Earth's Core

Cobalt is usually found in compounds with other substances and is rarely found in a pure state in nature. However, there is a source for pure cobalt in nature: metallic meteorites. These metallic meteorites are mostly iron, but they contain enough nickel and cobalt to turn the iron into an alloy of steel that is very hard. The steel in meteorites rusts very slowly.

Before falling to the surface of Earth, these meteorites were small meteoroids orbiting the sun in a region called the asteroid belt, which lies between the planets Mars and Jupiter. When the solar system was still forming, a planet similar to Earth broke apart. The core of this planet, like the core of Earth, was mostly iron with some nickel and cobalt. When this planet broke apart, it formed meteoroids made of an iron-nickel-cobalt

Mine Tailings

When a mine is being dug, not only is the desired ore removed, but many tons of unwanted rock are moved as well. When this unwanted rock, called tailings, is piled up near the mine, it can cause problems. Rainwater running through the broken rock can dissolve minerals that poison the water. When a mine is run responsibly, the tailings are treated to remove the poisonous substances. This is good for the environment, and it can produce useful by-products, too!

When meteorites fall to Earth, people get to handle an actual piece of rock from outer space. Metallic asteroids such as this one, the Manitou Stone, are usually an alloy of iron, nickel, and cobalt. This metal was harder than any that blacksmiths could forge until the eighteenth century.

alloy. Sometimes, one of these meteoroids deviates from its orbit and falls to the surface of Earth.

Cobalt in Industry

Many metals bend and melt when heated. But in an engine, the parts must not bend out of shape when the engine gets hot. If a little cobalt and nickel are added to the iron before it is cast into shape, the alloy is thermally resistant—the casting keeps its shape when heated. These steel alloys, called superalloys, have many high-temperature uses. These uses include restaurant cooking appliances, industrial furnaces, retorts (vessels where substances are distilled by heat), turbines, and heat-treatment fixtures. Coballoy is an alloy of tungsten carbide and cobalt used for making high-speed, hard-cutting tools used in cutting steel. Another alloy used to cut and machine steel is stellite, made from cobalt and chromium.

Some cobalt alloys are used for electroplating shiny metallic surfaces that don't tarnish. Superalloys resist corrosion caused by the intrusion of atoms such as oxygen, carbon, sulfur, or nitrogen into the metal surface. This makes superalloys valuable for use in waste incinerators and also in pumps, valves, and fittings for oil and gas well piping, or handling chemicals and radioactive waste.

Superalloys make good seals with glass or ceramics because they expand in the same way as glass when they are heated. They are also used in transistors, bimetallic strips for thermometers, electric transformers, and relay components for high-sensitivity ground fault circuit breakers.

Steel from the Skies

Steel was invented less than three thousand years ago. So, how could archaeologists find a steel knife in a four thousand-year-old Egyptian pharaoh's tomb? And how did Inuit people from Greenland and Canada's far north have a few steel knives, even though they didn't smelt any metals? The answer is metallic meteorites. These people found metallic meteorites that had fallen from the sky. The metal of meteorites is usually an alloy of iron, nickel, and cobalt, which is a form of steel. Pieces could be broken off to make useful tools.

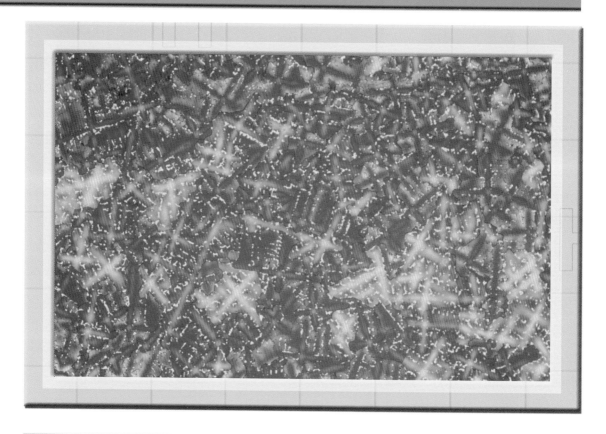

This photograph was taken from a microscope. It shows a polished sliver of a superalloy used for the turbine blades of a jet engine. The additives provide an interlocking structure that holds the alloy together. It's almost impossible to cut such an alloy into shape, so it is usually molded instead.

They are used for liquefied gas storage and for making composite wing molds for airplanes.

Cobalt in Catalysts

Catalysts promote chemical reactions among other substances but aren't used up in the reaction. Cobalt is used with molybdenum (Mo) as a catalyst to take the sulfur out of crude oil, which produces cleaner fuels. The original catalyst for petroleum refinement was a cobalt compound, $HCo(CO)_4$, but now platinum (Pt) catalysts are used as well. Cobalt catalysts help turn petroleum into many practical products such as the phthalates that are used to keep polyvinyl chloride plastics soft and flexible. However, these phthalates should not be used in plastics used for food containers because they can leach out into the food and enter our bodies.

Catalytic converters in automobiles use a catalyst to remove troublesome compounds from exhaust. Most car engines have a catalytic converter to remove nitrogen oxides from the exhaust. Most catalytic converters use a platinum catalyst, but this is very expensive. New catalysts using cobalt are being developed. Cobalt is a less expensive material for catalytic converters, but it is just as effective.

Before 1920, cobalt was mostly used in compounds as a colorant for glass and ceramic materials. But more than 70 percent of cobalt today is utilized as a metal, usually in tools made of various alloys. According to the U.S. Geological Survey, about 46 percent of cobalt is currently used in superalloys (chiefly for aerospace uses), 9 percent in magnetic alloys, 9 percent in cemented carbides, 6 percent in other alloys including steel, and 30 percent in chemical and ceramic uses. The British Geological Survey estimates that more than 35,000 tons (32,000 metric tons) of refined cobalt were produced around the world in 2000.

Chapter Four
Cobalt Compounds

Many cobalt compounds can dissolve in water. When they do, the compounds come apart into ions, which are atoms or groups of atoms that have an electric charge.

Cobalt atoms are able to lose either two or three electrons from their outer shells. When a cobalt ion loses two electrons, it becomes the ion Co^{2+}. This ion combines with negative ions to form a cobalt(II) compound such as cobalt(II) chloride, $CoCl_2$. When a cobalt atom loses three electrons, it becomes the ion Co^{3+}. It combines with negative ions to form a cobalt(III) compound, like cobalt(III) chloride, $CoCl_3$. Cobalt(II) compounds are common in rocky ores. Besides simple salts, cobalt ions also form complex compounds such as ammines that result when cobalt salts are combined with ammonia. Some of these complex compounds have industrial uses as catalysts. There are also compounds of cobalt(I), cobalt(IV), and cobalt(V), but these are less stable and seldom found in nature.

Cobalt as Coloring Agent

Since ancient times, people noticed that if some cobalt ores were soaked in water, the water would turn blue. People who washed clothes paid attention. Linen cloth is creamy white in color when newly made, but gradually over a year or two, it turns yellowish. People want their clothes to look clean

This paper is coated with cobalt chloride. Add a drop of water, and the cobalt compound attaches two molecules of water, which changes the color to pink. Heat the paper to dry it, and the color changes back to blue.

as well as be clean. If linen or cotton cloth is soaked and washed in water stained blue, the cloth looks whiter. Since ancient times, cobalt compounds have been used as laundry bluing.

Cobalt in Glass

Cobalt compounds have been used to color glass since the time of the ancient Egyptians. In the tomb of the Pharaoh Tutankhamen, there are pieces of cobalt blue glass. First, silica from beach sand is melted to make glass. If cobalt oxide is added, the glass turns a deep royal blue.

In 1777, Swedish chemist Johan Gottlieb Gahn and German chemist Carl Friedrich Wenzel were experimenting with a soldering blowpipe and learned that if cobalt aluminate is added, the molten glass turns turquoise. Cobalt silicate makes the glass a violet blue color. Cobalt oxide can also be added to a borosilicate glass, with boron added to the silica. The resulting color is purple or red, like a ruby or cranberry glass.

Only a small amount of cobalt—just 5 ounces (140 grams)—is needed to turn a ton of glass blue. But if a greater concentration of cobalt is added, the color will be darker. If 10 pounds (4.5 kilograms) of cobalt is added to that ton of glass, this creates a particularly dark glass called smalt. You can't use smalt to make decorative windows. Smalt is ground to make powdered ingredients for pottery glazes and enamels.

Cobalt in Ceramics

Pottery glaze is made by grinding smalt or minerals to powder, mixing the powder with water, and painting the clay pot. When heated in a kiln until the glaze melts, the surface of the clay is covered with a shiny finish. Cobalt blue is prized for its bright, deep color. Cobalt compounds have been used this way for hundreds of years in China. If you've seen a china plate with the renowned Blue Willow pattern, you've enjoyed a practical and decorative use of cobalt. By 1765, German potters were using what they then knew were cobalt oxides.

Cobalt in Enamels

Smalt is used when preparing blue colors for enamel work. Powdered minerals and glass are placed on metals and heated in a kiln. While some copper compounds can be used to make blue enamels, glass, or glazes, it takes a lot less cobalt to make a deep blue color.

The funeral mask of the ancient Egyptian pharaoh Tutankhamen and other orna-ments from his tomb are made of blue enamel work. Some of these fine pieces, and blue glass as well, were made more than 3,000 years ago using cobalt for color.

Cobalt in Paint

Artists have always used plants and rocks for color, mixing these pigments with water or oil or egg whites. It was hard to make blue paint that wouldn't fade. The best blue was made by grinding lapis lazuli, but this stone is rare and valued for jewelry.

Louis-Jacques Thénard was appointed by the French government to improve the colors used by artists. In 1802, Thénard announced that he had created a blue as beautiful as lapis lazuli, very stable, but costly at first. He ground 1 gram (0.04 ounce) of cobalt(II) oxide with 5 grams (0.2 ounce) of aluminum oxide, then heated the powder for three to four minutes. Cobalt blue, also called Thénard's blue or Dresden blue, was the first of the modern pigments that artists such as French painter Pierre-Auguste Renoir began using. By 1860, cobalt(II) stannate was being used to make cerulean blue. Other cobalt compounds made cobalt green and cobalt yellow. Cobalt compounds are used commercially today to make many inks, paints, and varnishes.

Minerals Needed by Animals

Sometimes, doctors prescribe vitamin and mineral supplements

> In his paintings, French artist Pierre-Auguste Renoir was pleased to use the new pigments being invented during the late nineteenth century. He painted *The Swing* in 1876, using blue from cobalt compounds.

Animal Medicines

Veterinarians use cobalt, in the form of cobalt chloride, cobalt sulfate, cobalt acetate, or cobalt nitrate, as animal dietary supplements and as medicine for correcting some mineral deficiency diseases in animals. As an element in the diet of sheep, cobalt compounds treat and can even prevent a disease called swayback—and it also improves the quality of the wool.

for people to take in addition to a good diet with a variety of fresh foods. Not only people but animals need minerals and vitamins, too. Farm animals deserve to live healthy lives, and when they are healthy, their meat, milk, and wool are a much better quality. Farm animals eat plants grown in farm fields. The soil should contain 0.13 to 0.30 parts per million of cobalt for proper animal nutrition. When there is not enough cobalt or other minerals in the local soil, farmers put out salt blocks containing mineral supplements. The animals need a little salt in their diet just like people do, and they enjoy licking the block as much as you enjoy eating a salty snack. Plain salt is white, but a block with cobalt and other minerals added is blue or reddish in color.

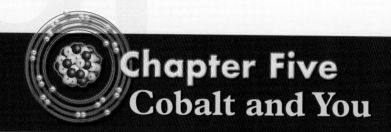

Chapter Five
Cobalt and You

Your body contains a small amount of cobalt. The body of a person who weighs 160 pounds, or has a mass of 75 kilograms, contains about 3 milligrams of cobalt. For every billion atoms in a human body, only about four atoms are cobalt. That's not much, compared to the other elements found in the human body, such as oxygen and carbon. But it's enough to supply the catalysts needed for several important chemical reactions. In every cell of your body, there are about 200,000 atoms of cobalt.

Cobalt is an essential trace element in your diet. A necessary part of your diet is vitamin B_{12}, which is a compound of cobalt. The recommended daily allowance of vitamin B_{12} is 2.4 mg. This vitamin is active in the production of red blood cells and in the formation of DNA. Cobalt is also involved in the biochemistry of methane-producing bacteria that live in composting soil and in animal intestines.

Cobalt in Artificial Joints

Do you know someone who has had a joint replaced? Doctors in North America usually make an artificial hip joint or a knee joint cast out of a titanium alloy. In Europe, cobalt chrome alloys are widely used, not titanium.

Artificial joints, such as this artificial hip that was made from a cobalt alloy, can help a person be active instead of unable to move around. Any prosthesis, which is what an artificial body part is called, that is used in the body must last for a long time without rusting or reacting.

And instead of casting, a cobalt chrome alloy part can be built in layers from metal powder. Each layer is formed when an electron beam (controlled by a computer-assisted drawing pattern) melts the metal powder. Then another layer of metal powder is laid and melted. This is done in a vacuum. The finished piece is solid with no air bubbles or oxidation. This new electron beam vacuum melting process is faster and easier than casting or carving the same shape. It's also useful for making strong, corrosion-resistant parts for the aerospace industry.

Cobalt in Our Cities and Technology

You may see a lump of pure cobalt only in a chemistry display. Yet cobalt is a small but important part of the technology that keeps our cities working. Cobalt is essential as a catalyst in petroleum refinement. From oil and gas piping to pollution control equipment, the incinerators that burn your community's trash, and even commercial pickle-making factories—all have parts that are made with corrosion-resistant alloys of steel using cobalt and nickel.

During a computerized axial tomography (CAT) scan, the soft tissues of a person's body show on the screen because a small amount of radioactive cobalt-60 is injected into the blood. Images of the cross sections of the anatomy can be assembled into a 3-D image to help doctors come to a diagnosis and develop treatment options.

Most forms of transportation use cobalt. Gasoline engines for cars and buses have catalytic converters to clean the exhaust. Jet engines for airplanes, and the molds to make airplane wings, use cobalt alloys. Even parts of skateboards and bicycles may contain cobalt alloys.

Radioactive isotopes of cobalt can be used for industrial purposes or medical imaging and treatment. Factories and water treatment systems track leaky pipes with radioactive dye. If you or someone you know has been treated for cancer with radiation treatments, the radiation source was probably cobalt. If you've had a CAT scan, the radioactive dye used to trace out your blood vessels probably had cobalt in it.

Cobalt is no longer an unwanted problem. It has become so irreplaceable for making superalloys and high-performance magnets that the U.S. Defense Logistics Agency, which buys, stockpiles, and disposes of vital raw materials for all branches of the U.S. armed forces, has stockpiled strategic reserves of this now valuable resource.

The Periodic Table of Elements

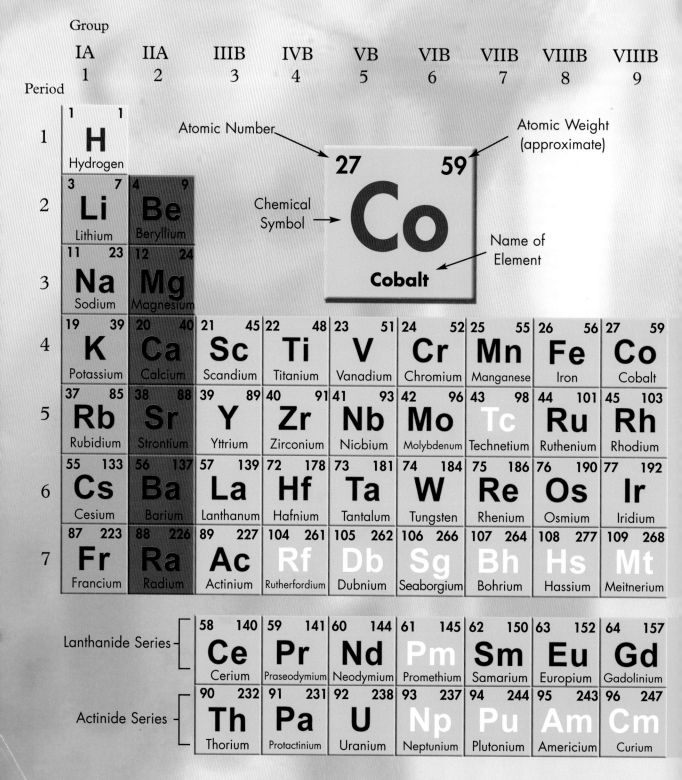

Group								
IA 1	IIA 2	IIIB 3	IVB 4	VB 5	VIB 6	VIIB 7	VIIIB 8	VIIIB 9

Period

Atomic Number

Atomic Weight (approximate)

27 59 Co Cobalt

Chemical Symbol → Co

Name of Element

| 1 | 1 1 **H** Hydrogen | | | | | | | | |

| 2 | 3 7 **Li** Lithium | 4 9 **Be** Beryllium |

| 3 | 11 23 **Na** Sodium | 12 24 **Mg** Magnesium |

| 4 | 19 39 **K** Potassium | 20 40 **Ca** Calcium | 21 45 **Sc** Scandium | 22 48 **Ti** Titanium | 23 51 **V** Vanadium | 24 52 **Cr** Chromium | 25 55 **Mn** Manganese | 26 56 **Fe** Iron | 27 59 **Co** Cobalt |

| 5 | 37 85 **Rb** Rubidium | 38 88 **Sr** Strontium | 39 89 **Y** Yttrium | 40 91 **Zr** Zirconium | 41 93 **Nb** Niobium | 42 96 **Mo** Molybdenum | 43 98 **Tc** Technetium | 44 101 **Ru** Ruthenium | 45 103 **Rh** Rhodium |

| 6 | 55 133 **Cs** Cesium | 56 137 **Ba** Barium | 57 139 **La** Lanthanum | 72 178 **Hf** Hafnium | 73 181 **Ta** Tantalum | 74 184 **W** Tungsten | 75 186 **Re** Rhenium | 76 190 **Os** Osmium | 77 192 **Ir** Iridium |

| 7 | 87 223 **Fr** Francium | 88 226 **Ra** Radium | 89 227 **Ac** Actinium | 104 261 **Rf** Rutherfordium | 105 262 **Db** Dubnium | 106 266 **Sg** Seaborgium | 107 264 **Bh** Bohrium | 108 277 **Hs** Hassium | 109 268 **Mt** Meitnerium |

Lanthanide Series

| 58 140 **Ce** Cerium | 59 141 **Pr** Praseodymium | 60 144 **Nd** Neodymium | 61 145 **Pm** Promethium | 62 150 **Sm** Samarium | 63 152 **Eu** Europium | 64 157 **Gd** Gadolinium |

Actinide Series

| 90 232 **Th** Thorium | 91 231 **Pa** Protactinium | 92 238 **U** Uranium | 93 237 **Np** Neptunium | 94 244 **Pu** Plutonium | 95 243 **Am** Americium | 96 247 **Cm** Curium |

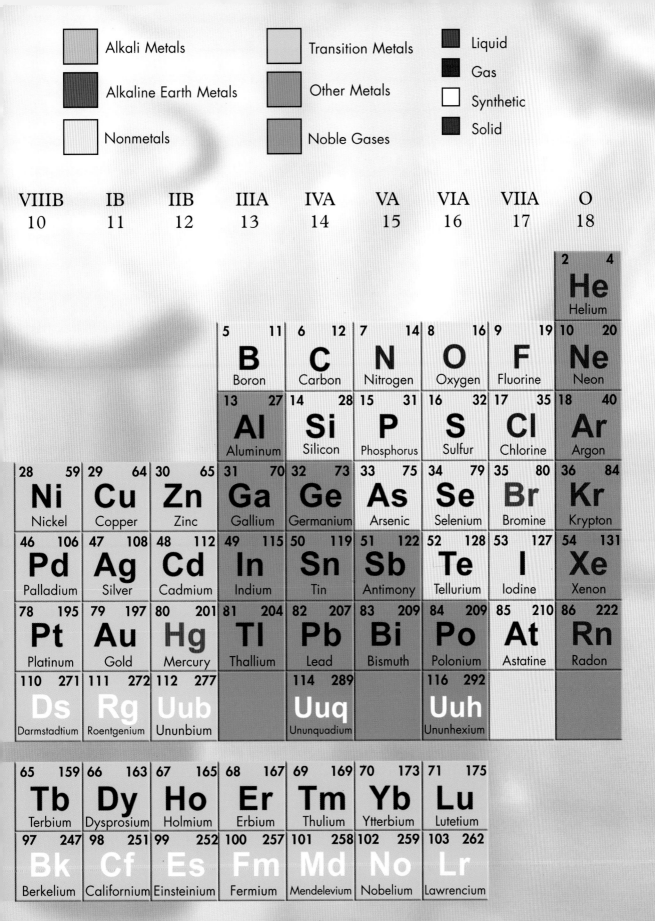

Alkali Metals · Transition Metals · Liquid
Alkaline Earth Metals · Other Metals · Gas
Nonmetals · Noble Gases · Synthetic · Solid

VIIIB 10	IB 11	IIB 12	IIIA 13	IVA 14	VA 15	VIA 16	VIIA 17	O 18
								2 4 **He** Helium
			5 11 **B** Boron	6 12 **C** Carbon	7 14 **N** Nitrogen	8 16 **O** Oxygen	9 19 **F** Fluorine	10 20 **Ne** Neon
			13 27 **Al** Aluminum	14 28 **Si** Silicon	15 31 **P** Phosphorus	16 32 **S** Sulfur	17 35 **Cl** Chlorine	18 40 **Ar** Argon
28 59 **Ni** Nickel	29 64 **Cu** Copper	30 65 **Zn** Zinc	31 70 **Ga** Gallium	32 73 **Ge** Germanium	33 75 **As** Arsenic	34 79 **Se** Selenium	35 80 **Br** Bromine	36 84 **Kr** Krypton
46 106 **Pd** Palladium	47 108 **Ag** Silver	48 112 **Cd** Cadmium	49 115 **In** Indium	50 119 **Sn** Tin	51 122 **Sb** Antimony	52 128 **Te** Tellurium	53 127 **I** Iodine	54 131 **Xe** Xenon
78 195 **Pt** Platinum	79 197 **Au** Gold	80 201 **Hg** Mercury	81 204 **Tl** Thallium	82 207 **Pb** Lead	83 209 **Bi** Bismuth	84 209 **Po** Polonium	85 210 **At** Astatine	86 222 **Rn** Radon
110 271 **Ds** Darmstadtium	111 272 **Rg** Roentgenium	112 277 **Uub** Ununbium		114 289 **Uuq** Ununquadium		116 292 **Uuh** Ununhexium		

65 159 **Tb** Terbium	66 163 **Dy** Dysprosium	67 165 **Ho** Holmium	68 167 **Er** Erbium	69 169 **Tm** Thulium	70 173 **Yb** Ytterbium	71 175 **Lu** Lutetium
97 247 **Bk** Berkelium	98 251 **Cf** Californium	99 252 **Es** Einsteinium	100 257 **Fm** Fermium	101 258 **Md** Mendelevium	102 259 **No** Nobelium	103 262 **Lr** Lawrencium

Glossary

atomic number The number of protons in the nucleus of an atom of an element.

atomic weight Also known as atomic mass. The average of the masses of all the different naturally occurring isotopes of an atom of a specific element.

ceramic Clay that has been fired in a kiln until it is as hard as stone.

cobaltite A cobalt ore containing cobalt, arsenic, and sulfur.

compound A substance containing the atoms of two or more elements joined together by chemical bonds.

density The amount of mass an object has in a given volume, often expressed as grams per cubic centimeter (g/cm^3).

group In the periodic table, each vertical column of elements.

ion An atom or molecule with an electric charge due to the loss or gain of electrons.

isotopes Atoms that have the same number of protons but a different number of neutrons. Isotopes are different forms of a particular element.

mass number The total number of protons and neutrons in the nucleus of an atom.

metalloid An element that has some properties of a nonmetal and some of a metal.

oxidation The process of combining with oxygen or with another reactive nonmetal.

period In the periodic table, each horizontal row of elements.

valence electrons The outer shell of electrons that allows atoms to link together chemically and metals to conduct heat and electricity.

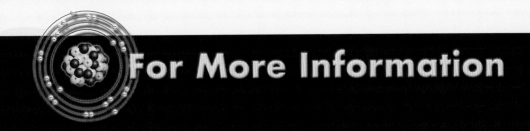

International Union of Pure and Applied Chemistry
IUPAC Secretariat
P.O. Box 13757
Research Triangle Park, NC 27709-3757
(919) 485-8700
Web site: http://www.iupac.org
IUPAC is an international body designed to advance the chemical sciences.
Science Across the World (www.scienceacross.org) and Young
Ambassadors for Chemistry (YAC) are just a few of the educational
programs you will discover on its Web site.

Jefferson Lab
12000 Jefferson Avenue
Newport News, VA 23606
(757) 269-7100
Web site: http://education.jlab.org/itselemental/ele027.html
Thomas Jefferson National Accelerator Facility (Jefferson Lab) is a
research laboratory that conducts basic research of the atom's
nucleus at the quark level. Its Web site features homework help and
online games and puzzles for students and resources for teachers.

Los Alamos National Laboratory
P.O. Box 1663
Los Alamos, NM 87545
(888) 841-8256
Web site: http://periodic.lanl.gov

Los Alamos National Laboratory's Chemistry Division maintains a Web resource on information about the periodic table of the elements, including element properties and sources.

Oregon Museum of Science and Industry
1945 South East Water Avenue
Portland, OR 97214-3354
(503) 797-4000
Web site: http://www.omsi.edu
This museum offers online information about the periodic table of elements and a guide to resources.

Web Sites

Due to the changing nature of Internet links, Rosen Publishing has developed an online list of Web sites related to the subject of this book. This site is updated regularly. Please use this link to access the list:

http://www.rosenlinks.com/uept/coba

For Further Reading

Emsley, John. *Nature's Building Blocks: An A–Z Guide to the Elements.* New York, NY: Oxford University Press, 2003.

Gonick, Larry, and Craig Criddle. *The Cartoon Guide to Chemistry.* New York, NY: HarperCollins, 2005.

Hudson, John. *The History of Chemistry.* New York, NY: Routledge, 1992.

Knapp, Brian J. *The Periodic Table.* Danbury, CT: Grolier Educational, 1998.

Oxlade, Chris. *States of Matter* (Chemicals in Action). *2nd ed.* Chicago, IL: Heinemann Library, 2007.

Saunders, Nigel. *Gold and the Elements of Group 8 to 12* (Periodic Table). Chicago, IL: Heinemann Library, 2003.

Stwertka, Albert. *A Guide to the Elements.* 2nd ed. New York, NY: Oxford University Press, 2002.

Watt, Susan. *Cobalt* (The Elements). New York, NY: Benchmark Books, 2006.

Bibliography

Bentor, Yinon. "Cobalt." ChemicalElements.com. Retrieved May 23, 2007 (http://www.chemicalelements.com/elements/co.html).

Cotton, F. Albert, and Geoffrey Wilkinson. *Advanced Inorganic Chemistry.* 6th ed. New York, NY: John Wiley & Sons, 1999.

ETC Group. *The Big Down: From Genomes to Atoms. Atomtech, Technologies Converging at the Nano-Scale.* Winnipeg, Manitoba, Canada. January 2003. Retrieved May 2, 2007 (http://www.etcgroup.org).

Gonick, Larry, and Craig Criddle. *The Cartoon Guide to Chemistry.* New York, NY: HarperCollins, 2005.

Stwertka, Albert. *A Guide to the Elements.* 2nd ed. New York, NY: Oxford University Press, 2002.

Thomas Jefferson National Accelerator Facility—Office of Science Education. "It's Elemental—Cobalt." Retrieved May 23, 2007 (http://education.jlab.org/itselemental/ele027.html).

Tweed, Matt. *Essential Elements: Atoms, Quarks, and the Periodic Table.* New York, NY: Walker & Co., 2003.

Index

About the Author

For twenty years, Paula Johanson has worked as a writer and teacher. She writes and edits nonfiction books. At two or more conferences each year, she leads panel discussions on practical science (usually biochemistry) and how it applies to home life and creative work. Her pottery and clay figures are glazed with cobalt compounds, and her kayak is stained cobalt blue. An accredited teacher, she has written and edited curriculum educational materials for the Alberta Distance Learning Centre in Canada.

Photo Credits

Cover, pp. 8, 9, 11, 40–41 Tahara Anderson; pp. 5, 37 © SSPL/The Image Works; p. 7 © Trine Thorsen/Red Cover/The Image Works; p. 17 © Jacques Boyer/Roger-Viollet/The Image Works; p. 18 © Dick Blume/Syracuse Newspapers/The Image Works; p. 20 © Roger Ressmeyer/Corbis; p. 23 © Per-Anders Pettersson/Getty Images; p. 24 © The Natural History Museum/Alamy; p. 26 Courtesy of the Royal Alberta Museum, Edmonton, Alberta; p. 28 © G. Muller, Struers GMBH/Science Photo Library/Photo Researchers, Inc.; p. 31 © Andrew Lambert Photography/Science Photo Library/Photo Researchers, Inc.; p. 33 © Werner Foreman/Topham/The Image Works; p. 34 © Réunion des Musées Nationaux/Art Resource, NY; p. 38 © Tony Savino/The Image Works.

Designer: Tahara Anderson; **Editor:** Kathy Kuhtz Campbell
Photo Researcher: Amy Feinberg